PREDATOR!

KNOWING NATURE®

PREDATOR!

Bruce Brooks

Farrar Straus Giroux · New York

In association with Thirteen/WNET

For Alexander

Library of Congress Cataloging-in-Publication Data

Brooks, Bruce.

Predator! / Bruce Brooks. — 1st ed.

p. cm.

Includes index.

Summary: Survival in the wild creates a hierarchy of predators and their prey; this interaction among the animals forms the basis of a complex ecological system known as the food chain.

1. Predation (Biology)—Juvenile literature. [1. Predatory animals. 2. Animals —Habits and behavior.] I. Title.

QL758.B76 1991 591.53—dc20 91-3369 CIP

Contents

PREDATOR!

Food

Whenever you see a wild animal of any kind—from a fox trotting through the woods to a fly zipping around your bedroom—the chances are it's looking for something to eat. While we human beings have the luxury of spending our days at the things we choose—at work and play—almost every snake, bug, bird, fish, frog, and mouse spends much of its waking time trying to get enough food to stay alive for the next hour or day. Then the next hour or day is spent looking for *more* food, to stay alive for *another* day, which will also be spent looking . . .

The simple fact is, for a wild animal the future is hunger. This means the present cannot be settled into, with any kind of relaxed security—a belly that is full today will be empty tomorrow. The energy that is gained from eating a good meal must be used up in hunting for the next meal. There are some breaks from the cycle of eating and searching—migration and mating season, for example. But an animal pays for these breaks with heavier food-gathering duties: the long trip of migration requires a buildup of fat gained by extra eating in advance; and when pups or chicks are born, they must be fed tremendous amounts of food. The parents of baby

Termites are apparently one of the world's fine foods—many larger creatures live on them entirely. This pangolin, ripping into a termite mound with claws and tongue designed for the job, may eat more than 300,000 in one day.

birds will make hundreds of trips a day to leave the nest, find a caterpillar or two, and bring them back to a thicket of screaming mouths.

There are some animals that pretty much live their whole lives directly *on* their food source—bugs that are born on a plant, spend every day eating the plant's leaves, lay their own eggs on the plant, and die there, too. A great many of the world's animals, however, have a life-style in which the search for food is more interesting, if less secure. They must go look for the thing they like to eat, and when they find it, they kill it. This is because what many of the world's creatures like to eat best is other animals.

An animal (or plant) that kills another animal for food (or, more accurately, that kills a *lot* of other animals for food in the course of a lifetime) is called a predator. Most predators get to run around quite a bit more than the on-site leaf-chewers described above, but they are no less dominated by the need to find food, and the task is considerably more difficult when your food can run away from you, or fight back.

A predator's life seems to have all the marks of thrilling strength, lonesome danger, and intimate desperation. Though a few kinds of ants and fish and birds and mammals hunt together, in most species a creature is on its own—often from birth—in the unending competition for food. When an owl is floating beneath a cloudy moon, passing field after field, stream after stream, with no sign of a small critter to swoop upon, he is certainly a lonesome figure. But if he does discover something to kill—a shrew that scoots quickly, but not quickly enough, along the banks of a stream—the singular act of death that follows seems violent, almost personal: practically murder. It is one big, strong creature snapping the neck of a weaker one and gobbling him up. Gross. Cruel.

Beetles, despite the beauty of their design, are valued more as food than as sculpture.

But when we look at the stream a little more closely, and see that the shrew was scooting around his bank looking for frogs, which were themselves swimming after minnows, which were cruising beneath the surface hoping to spy some dragonfly larvae, which were busy gobbling up mosquito larvae . . . Well, then we begin to see the whole picture. We begin to see that the eating of smaller things by larger things is a kind of system—that everyone is chasing someone else, and that far from being a lot of random violence, these acts of death establish a remarkable balance in nature. The owl's solitary quest does not seem so pitiful anymore, nor his predation so cruel.

This system of predation is called the food chain. Like the links in a chain, the acts of eating are fastened one to the next, and the whole thing works only if each part functions in strength. Essentially, the fate of the owl is dependent on the mosquito larva, because only if each smaller creature finds what it needs to eat will the next-larger creature find what *it* needs, and so on. If one goes hungry and dies out, all are affected.

The nasty assassin bug—it kills, it stinks, it carries disease. Some birds, however, find it tasty.

This white-fringed antwren doesn't look like much of a predator, unless you are an insect.

The genet is a prowler of rigid habit: it roams the same path at the same time, day after day, hoping for chance to place a large bird or small mammal in its way. (Its eyes, by the way, don't always glow green—here a layer of reflective cells behind the retina has caught the photographer's light source.)

The food chain exists in all environments of the earth. All predators have a part in their local system. But, naturally, the system has to start with something that does not prey upon anything else—something that does not eat but is eaten. In perhaps the most complex food chain—that of the sea—the beginning is microscopic algae. Using light as a power source, these plants turn air and water into food. The algae are eaten by one-celled protozoans (very simple, microscopic animals), which are eaten by tiny crustaceans (such as a primitive shrimp called krill), which are eaten by small fish and octopuses, which are eaten by larger fish, which are eaten by barracuda or sharks.

Thus the food chain starts with something that does not eat but is eaten; it ends with something (the barracuda or shark in the sea; the tiger or alligator on land) that eats but is not eaten. This fortunate, dashing survivor is said to be at the top of the food chain, while the microscopic tidbits, which any blind, wimpy, newborn, or slow critter can capture, are said to be at the bottom.

For all of the seeming neatness of its small-to-large steps, the food chain is terribly inefficient in some ways. Every time a creature is eaten, much of its residual energy is not converted to new energy in the predator but is lost as waste. If a Caribbean fisherman catches the barracuda in the food chain above and eats it, the whole conversion may have used ten thousand pounds of sea microorganisms to build one pound of fisherman. That is not a great ratio. It involves a lot of death, too, and a lot of energy

Microscopic marine animals, with minute plants, make up a stew called plankton that is the "first food" in the earth's oceans.

The Nile crocodile is the ultimate predator at the top of the food chain: who is going to try to eat a twenty-foot-long steel-skinned lizard with a mouth you could park a small car in? This one is using its mouth for a better task—carrying one of its babies.

wasted hunting and hiding. Wouldn't it be easier if nature simply provided some sort of Superfood that was available to all creatures?

Surprisingly, the large predators would probably be the ones to prefer such a simple chow arrangement. Despite the appearance of domination, the predatory animals at the top of the chain are rarer, less adaptive, more restricted in diet, and more vulnerable to environmental changes than the creatures further down. A killer whale or leopard depends on so many prey-predator relationships remaining intact that the slightest disturbance can topple him. When something goes wrong and a local insect population fails, the insects may survive in temporarily dwindled numbers, as may the bird population that eats them. But as the dwindling is passed upward, the population of the top predator in the area, already represented by far fewer animals than the bugs or birds, may not survive the same temporary drop in numbers. In other words, the dwindling may turn out to be permanent for this otherwise omnipotent animal.

But there is no Superfood, no full trough where each rabbit and beetle and bat and lobster may go a few times a day to stoke up. This would make for a fairly static world—a world, perhaps, such as human society,

Don't try to get away from the Siberian tiger by jumping in the water. It welcomes the chance to swim for its supper.

where we are all one species with very little biological diversity. Instead, nature cast us all into the middle of a wild drama full of the most extreme kinds of action, which give rise to the richness of varied life forms all over the planet.

For it is the quest for food—and the opposite quest, to avoid being eaten—that leads each species to the next step of progress in its evolution. To gain an edge in the struggle for flying insects, the nighthawk develops a wide mouth surrounded by bristles that work like a sweeping net; to keep from being gobbled up by every halibut that lumbers along, the squid comes up with ink to squirt in the face of the curious or hungry. Even we humans owe our most critical inventions to the fight for food: the shaped stone and wood weapons and tools that made us suddenly different from the beasts around us in the prehistoric scramble.

A look at predators is most exciting perhaps because the quest for food shows every creature on earth at its best. You must use every tool, every bit of intelligence and speed and strength, when the one thing you do is a matter of life or death. We humans have the privilege of forgetting this; we can skip a bowl of Grape-Nuts in the morning and survive just fine. When we watch animals, however, some innate appreciation of the endless drama is awakened. Once again we identify with the keen, aggressive strategy of the hunter, with the shrewd, desperate feints of the prey. We stand outside the struggle, but our heart is in it.

The large, furry, padded paws of the lynx give it good footing for snowy chases of prey.

Hunting

To eat something, you first have to find it. Most animals have an aversion to being eaten, and they try to avoid putting themselves in the position of easy prey. They hide, they flee, they disguise themselves— they use all kinds of defensive strategies to make it as hard as possible for predators to get at them.

So a predator has work to do before it can eat. There are as many ways to find prey as there are places to hide, but most predators operate by one of two main methods. One is to look for the prey animal where *it* lives or hides. The other is to wait until it happens to come by where the predator lives or hides. We could call the predators who go out and seek prey the prowlers; the others we could call the ambushers. In both groups there are animals who use great genius and cunning (plus a lot of precious energy), as well as animals who sort of loaf around and count on good luck.

It is easy to see that many predatory animals are smart and gutsy, and that many also possess what we might call honor. They think about where to find their prey; they plan how to attack or make their prey come to them; they confront the prey (sometimes they are pretty sneaky about it) and fight; they kill what they will eat, and usually no more. In its way, this seems a simple kind of life, full of stark conflicts—strength against strength, with the risk of defeat balanced against the glory of triumph.

But even the noblest hunting animals don't think much about glory. Mostly they think about hunger. For each species, a history of centuries spent hunting, generation after generation, has refined instincts and equipment, so a starfish or a killer whale knows from a young age exactly what it will have to do if it wants to eat.

Sometimes the killing starts even before the animal has come into its adult body. In many species of insect, the deadliest form is the larva—the squiggly wormish stage between egg and adult shape. Some larvae, such as those of several wasps, kill in the most innocent way possible: they hatch from their eggs and simply begin to eat the food that is around them. What makes them killers, however, is that the food around them is the living tissue of a spider, beetle, or caterpillar, because their mother inserted her eggs beneath the skin of this host. The eggs ride along, snug and well nurtured; the young are born inside the host; they eat, starting (by amazingly precise instinct) with the least vital parts of the insect, so it will stay alive until they are strong enough to leave. Finally the hollowed-out bug yields its last drop of nutrients, and the young wasps abandon it like a Coke can, to strike out on their own. The timing is perfect, refined to true science by thousands of years of practice.

Other larvae are not born so nicely placed and have to come up with more aggressive hunting behavior. The ant lion larva is one of the most talented of these—a fearsome hunter in the nightmares of lesser insects.

If you saw an adult ant lion, you would wonder why it had such a goofy name: obviously this beautiful, slender dragonfly look-alike, with four graceful wings, has nothing ferocious or powerful about it. Obviously, too, it doesn't bother ants—its mouth is too small and innocuous and it turns out to be a weak, indifferent kind of flier, unable to pursue much of anything. The ant lion, a wisp in the air, flitting feebly to a leaf, looks harmless, even merely decorative.

The larva, however, looks as if it's right out of the next alien scare movie. It has the usual low-slung, wiggly larva body, and inadequate legs that seem barely able to keep its bulbous sac off the ground. But up front, on the head, loom a jagged set of pincers, clearly designed for sharp, crushing snatches. It's as if the little bug had somehow managed to swallow a much

The trapdoor spider lies snug in its hole until something walks overhead. Then it springs out like a nightmare come to life.

larger crab, and everything had gone down except one fierce claw left sticking out.

The ant lion larva cannot rely on its pincers alone; overpowering though they may look, they are unwieldy and overt. An ant could not fail to see the ant lion coming, and to read the pincers as a sign proclaiming WARNING! I WANT TO EAT YOU! So if the larva wants to profit from its weapon, it must embark on a construction project that is a perfect example of how animals will modify their environment to make the best use of their physical equipment. The larva spins its small body like a screw and buries itself, jaws up, in loose sand. The wide-set pincers are left outside, sprung open, like a bear trap, ready to snap shut on any edible insect that happens to stumble right between them.

The ant lion larva decided it couldn't depend wholly on time and luck, waiting for something tasty to wander by; it needed to improve its chances by artifice. So instead of just burrowing flush into the surface of the sand, it does some extra digging. It digs a cone-shaped hole, snapping its head upward to throw a very loose layer of grains onto the sloping sides. The

This ant lion larva is burrowing backward into the sand at the bottom of its pit, with its jaws already cocked.

larva ends up at the bottom of a kind of whirlpool pit. When an insect—most often an ant—steps on the edge of the pit, the loose sand slides. The insect scrambles, but it only dislodges more loose sand, and slips down the side of the pit toward the bottom. There it lands in the jaws, which snap shut. Using its mouth, which lies between the pincers, the ant lion larva opens a hole in the trapped bug and sucks it dry.

As effective as this is, the ant lion larva is a crude caveman next to the supreme artists of the ambushing animals: the spiders. Some spiders dig pits, but the finest and most deadly art of this group of predators is the silk web that sits almost invisibly in the path of flying insects and snares them in its sticky strands. (An invisible trap is bad enough—but some webs even lure the insects by reflecting ultraviolet light in patterns that make the webs look like flowers in the dark.) The spider comes along at its leisure, paralyzes the prey with a venomous bite, wraps it up, hauls it away, and later injects it with digestive fluids that turn its insides into easy-to-slurp goo. Some spiders, however, are energetic stalkers, pursuing game for long distances that they cover rapidly on their eight legs. One species

Perhaps the spiderweb, for catching flying insects, inspired fishermen to take to the sea with nets long, long ago.

of large and undeniably aggressive spider on the Argentine pampas can
keep pace with trotting horses; when one rider flicked his whip at such a
pursuer, the five-inch spider leaped upon the leather thong and scrambled
up the shaft. The rider dropped the whip pronto.

Amphibians are usually portrayed as classic ambushers, and it is true
that toads and frogs can be found squatting stone-faced and still in the
shallow edges of ponds, waiting until the right flying insect comes close
enough, and then—ZAP—the bug is snatched from the air by the famous
darting tongue. But most frogs and toads hunt on the prowl, skillfully
pursuing and catching all kinds of invertebrates, fish, reptiles, and small
mammals. When a bullfrog gets a mouse, it is because it chased it, cornered
it, caught it, subdued it, and—with that huge mouth—gulped it down.

Some animals ambush when they can, and do the chase-and-catch bit
when they must. A polar bear that spots a seal sunning itself will creep

*No, that's not the bullfrog's tongue—this adroit hunter has caught itself a
ribbon snake.*

on its belly for hundreds of yards across open ice, moving with an un-expected subtlety to blend in with the blazing white surroundings; some people say the bear will even keep a chip of ice in front of its black nose, pushing it as it goes along. Eventually it will be close enough to charge the seal. If the seal is slow or far from the water, the bear will kill it on the ice; if the seal is quick, it will slip into the water, where the bear's pursuit of the faster swimmer would almost certainly be futile.

But this stalking and fighting is hard work. So the bear prefers its ambush strategy: by sniffing around, it finds a breathing hole that a seal has dug from below in the floating ice (seals are mammals and must breathe periodically, preferably without leaving the safety of the water). The bear waits for the seal to come by and stick its nose up for a quick breath. With its huge forearm, the bear reaches down and smashes the skull of the very surprised swimmer, and hauls it up, neatly breaking whatever bones it must to squeeze the fat prey through the narrow hole. Put a man in the polar bear's place and substitute a fish for the seal (and a fishing line for the bear's claw), and you have a favorite winter sport in Minnesota—ice fishing.

Ice fishermen, polar bears, spiders, moray eels, snapping turtles, and almost any predator with a good sense of energy conservation will prefer lying in wait to running around, though they are ready to move when necessary. Some creatures do not have the choice. The animals that we collectively call coral are anchored in one spot in the sea by the calcified exoskeleton they secrete (unlike mammals and most vertebrates, corals, lobsters, beetles, scorpions, and other invertebrates—as well as turtles and trunkfish among the vertebrates—keep the soft stuff inside a hard exterior). The animals inside the stony coral structures reach out into the water to siphon and snare whatever tiny animals float by in the plankton.

But in an odd gotcha-back struggle taking place in some parts of the Pacific, the reef-building corals are being devoured at an alarming rate—exoskeletons and all—by huge numbers of a starfish called the crown of thorns, which crawls over the coral, projects its own stomach outside its body to cover the coral with digestive juices, then sucks up the whole mess like a wet-vac. The crown of thorns is one of the least-preyed-upon animals in the world, perhaps second only to the killer whale: its back is covered with toxic spikes that give a paralytic sting to anything touching them. The only time this starfish is vulnerable is when it is a tiny larva spinning around with the currents, in the plankton. At that time, one of its predators is the same coral it later eats. To judge from the rate at which reefs from

The crown of thorns starfish has as few enemies as the Nile crocodile—which is how it is getting away with sucking the life out of coral reefs around the world. A shellfish, the triton, was a major predatory control, eating the starfish when they were tiny and without venom; but humans have almost wiped out the triton because its shell makes a pretty trinket.

Japan to Australia are disappearing under the hordes of starfish, the coral animals are losing the battle despite getting an earlier start.

Another kind of stationary predator stays in the place it grows, but depends less than others on catching tidbits borne on the breeze. The carnivorous plants—pitcher plants, venus flytraps, sundews, and others—have found a way to bring their prey right into their "jaws," by offering sweet-smelling nectar to the unwitting forager. The pitcher plants drown them; the flytraps close their spiny leaves like a hand turning into a fist; the sundews hold the bugs by sheer stickiness.

Most of the time, these plants consume insects, which seems creepy but acceptable; the world is so full of plants we don't like to think of them as

Coral is not dead rock in the sea—it is a colony of soft-bodied creatures that secrete a hard outer skeleton and reach tentacles out to filter plankton from the water that flows by.

actually *eating* anything, but if they must eat, bugs are no big loss. However, there are species of pitcher plant that have ambitions: they gobble up lizards, frogs, and (shudder) even mice. (This is the only known instance of plants eating mammals, and most of us would agree that mice are a good place to draw the line.)

These plants—equipped with the chemicals necessary to digest meat, instead of making do with the usual photosynthetic products—live by themselves in places with such poor soil that most other vegetation cannot survive there. To get the nitrogen, phosphorus, and other nutrients that plants usually get from the soil, the carnivores must look away from their root systems and into the animal world above.

Perhaps the best double-ambush duo in nature is the tag team of a certain crab spider and a species of pitcher plant in Malaysia. This spider hides just inside the lip of the pitcher plant's bowl, hanging on to the sides with a few threads of its silk, well above the acidic but sweet-smelling fluid that fills the bottom of the bowl and attracts insects. When bugs fly in to investigate the nectar aroma, they are grabbed by the spider and eaten. When the spider has had its fill, it continues to catch the bugs, but drops them into the juice to nourish the plant (which *would* have caught them anyway, all on its own). The spider, though it doesn't improve the plant's life at all, at least does no harm, so the plant does not close its lid and try to kill it. Chances are the spider could survive longer than the plant if this happened, as spiders are capable of great fasts, whereas the plant would die sooner with its door shut and no insects coming in. So the plant puts up with its crafty guest, and the spider makes a living with perhaps the easiest predatory life-style in the world.

Other animals team up to make the most of their opportunities. Lions hunt together, splitting a group of four to six females into two groups (it is the females who feed the family; the male's job is protecting the territory). One group sneaks downwind of the prey—a herd of wildebeest, say. The other group approaches the herd from upwind, so that its scent reaches the wildebeest. The herd smells the lions and moves away; the lions step up their march, once they make sure the herd is on the hoof in the right direction; the wildebeest move faster; the lions come closer; the wildebeest panic and bolt—right into the waiting jaws of the well-rested downwind lions, who are soon joined in the feast by the attackers.

Killer whales hunt in groups, too (called by the cryptic collective noun "pods"). One would think the whales hardly need to team up: the orca is the most daunting killer in the world—big, fast, brilliantly intelligent, with nature's scariest-looking teeth set in a grin of well-deserved self-confidence.

But team up they do; being intelligent often means being cooperative. So when one killer whale hoists himself vertically above the water's surface and spots a pattern in the water that means a school of fish or herd of seals is passing, he communicates the target's data to his pod mates, by wailing and hooting and smacking the water with his fins. The pod fans out, encircling the prey, and closes in for lunch. Sometimes orcas hunt in smaller teams: two of them will taunt a large baleen whale by cruising alongside it, one to a side, nipping and nudging and poking, like two hoods in stock cars hassling a stern old gent riding in a Cadillac between them. The orcas are trying to exasperate and exhaust the whale until it opens its mouth—at which point one of them grabs its huge tongue and the two of them eat it on the spot.

Unlike the lioness, which is not an especially talented solo hunter, a

Hyenas hunt in packs, bringing down larger animals and eating them alive, but they do not share very graciously. The eagle nearby is hoping for leftovers.

killer whale has the stuff to go it alone (for example, lone orcas have been known to hurl themselves onto the edge of an ice floe containing a seal or penguin, holding the disc like a slanted floor upon which the helpless animal falls and rolls, backwards, to its grinning tormentor). And going it alone has always seemed to us the noble way of the true hunter. Anybody, we think, can gang up to kill a few seals or antelopes. Who has the wits and guts to figure out all the angles and make the kill without assistance?

Quite a few predators do, especially those hunters we might describe as opportunistic prowlers—the ones who roam their territory endlessly and monotonously, hopeful of the chance encounter with something they can eat. Most of the great birds of prey operate this way. The larger hawk species, called buteos, hover high in the sky, keeping a sharp lookout for the odd motion below that means a small animal is on the move in exposed territory, while the smaller hawks and falcons cover their plot of land methodically, passing back and forth in straight lanes over a field as if they were designing a geometric plan for plowing it. Owls tend to perch and watch, though they silently scout on the wing, too.

Sharks, which have to stay in constant motion to regulate their response to water pressure, are perhaps the ultimate prowler. A shark swims for literally its entire life (which may last several decades), always keenly alert

The peregrine falcon's deadliest feature is speed: with flight speed of up to 200 mph, the peregrine is probably the fastest animal on earth (or, more correctly, above it). Its keen vision spots prey from vast heights, and it falls like a missile for the strike.

As if a killer whale weren't dangerous enough on its own! Hunting together, these brilliant animals are much more cooperative than the hyenas: one of them will lunge onto this ice floe, tipping the seal off for the others to kill quickly.

to vibrations or smells in the water that mean a sea animal somewhere is injured, lost, frightened, or weakened in any other way. The shark is incredibly sensitive to such signals, interpreting the slightest change in the water's tenor, and zeroing in if the signs indicate a probable kill. On land, snakes—especially vipers in the vast stretches of the desert—roam and seek similar vibrations, smells, or waves of body heat in the air.

We could presume that solo hunters who prowl like this have learned in the course of evolution that carefully designing a strategic kill is not worth the energy it takes. We could also presume these animals don't have the intelligence for the job, either. Certainly the shark is not regarded as the genius of the sea, nor the rattlesnake of the land. Killer whales and wolves, which zoologists tell us *are* very bright, do indeed hunt with strategy when they hunt alone, but, as mentioned, they prefer teamwork to solitude.

An owl's talons give a killing squeeze.

So what about the complex solo killers—the cheetah who watches a herd for days, selecting exactly the right antelope from afar, which it isolates and practically hypnotizes with strange, brilliant feints and charges, or the ant-eating spider that wears a hollow ant body on top of its own, rearranges its legs to mimic ant ambulation, and insinuates itself securely into an anthill before quietly beginning to slaughter from within? *Are* they so smart, or do we just call them that because their schemes are clear to us, their reasoning in line with our own way of thinking ("Hey, that's a good idea—that's how *I* would do it!")? Or, perhaps, are these animals a little *too* smart, making a complicated game out of something that could be as simple as waiting quietly and biting hard?

We cannot know what motivates an animal to hunt by patient pursuit instead of simple force. We can say that some animals seem to love the thrill of the noble hunt, or that brilliant hunting tactics earn a male greater prestige within a pack and therefore impress females at mating time, or that cheetahs are artistic and thoughtful while hyenas are crude and mechanical. One thing, though, cannot be argued: an animal hunts the way it does because that is the best way to get fast, regular meals. Animals are not smart for the sake of being smart, or artistic, or noble. They are hungry. When they need to eat, they must rely on their senses and be guided by how their instincts say they can best use sight and smell and touch and taste and hearing within their specific environment, to find an animal to eat. But once they've found it, they still have to kill it—which is another whole story.

Killing

All right, you're a leopard and you've been very clever and you've cornered an eighty-pound baboon with three-inch fangs and frightened him into an adrenaline-stoked frenzy of hatred. Great! Now what, hotshot? Remember—you're supposed to be *happy* about this; you've worked hard to put yourself in this position.

What you do, if you are a leopard, is simple. You fake a step forward, inducing the baboon to make a do-or-die lunge at your eyes with those fangs. Then you pull back. The fangs click together a half-inch in front of your face, and before the stumbling primate can open his mouth again, you swing your right leg sharply from the shoulder and clout him in the side of the chest. He flies ten feet and lands in a tangle of broken ribs and crushed organs, and you heave a sigh: whew! Not that you ever doubted your superior strength, or your speed, or your craftiness. But all the same, those fangs, if they *did* get your eyes . . .

The sternest fact of a predator's life is this: if you are going to go around getting dangerous animals into a fight for their lives, you'd better have the stuff to put them on ice.

If evolution dictates that an animal needs to kill to stay alive, it's only fair that nature make that animal dangerous. This is why predators have been blessed with some of the most glorious physiques on earth. In any given habitat, the flashiest animals—with the hot speed, the wild strength, the awesome teeth or claws or toxic stings—will be the ones that eat other creatures. This is one reason we are always fascinated with predators, why, in a zoo, we walk right past the tapirs to ogle the jaguar, and skip the tortoises to shudder in front of the king cobra. We want to see the top-of-the-line models with the jazzy equipment—the Ferraris, not the sub-compact station wagons.

Of course, it needs to be said that nature makes no judgments based on melodramatic flash; she does not favor one creature over another because it is fun to watch. The world is full of vegetarians, many of them beautiful, intelligent, interesting, and quite strong in their evolving populations . . . Yeah, yeah. So go watch a camel chew grass; the rest of us will check out the Komodo dragon and the great white shark.

Lions wrestle before they bite.

Actually, we must be careful not to focus only on the larger monsters. The animals at the top of the predatory food chain are not necessarily the best endowed, just the biggest. Some of the grandest killing equipment is revealed only under a microscope, along with some of the cleverest killing tactics. A tiny flower-shaped water animal called the hydra has perhaps the most fearsome weapon: a poisoned harpoon called a nematocyst. This weapon deserves a complex, technical-sounding name, for it is a masterpiece of design. There are varying forms, but common to most is a springy, barbed spear coated with a nerve toxin, coiled inside a cell or sac. The retractable spear fires out of the sac with explosive force when an arm of hydra tissue senses (chemically or through touch) the presence of an edible organism nearby. And it's not just one spear, either—it's hundreds or thousands. The unfortunate prey is punctured and poisoned at the same time; the venom, which in some hydras is similar to the stuff that makes the wasp's barb sting, stuns and paralyzes the victim, and the wounds allow the hydra to suck its juices.

The microscopic hydra has snared a tiny shrimp, stunned it with a few hundred venomous barbs, and is now ready to eat.

It is the nematocyst that provides the sting of the sea anemone, the fire coral, and the jellyfish. The "jellyfish" we fear the most, the Portuguese man-of-war, is not a jellyfish at all but a small floating army of cooperative animals banded together to make the most of their individual skills in capturing, killing, and digesting prey. In this co-op, the marksmen, responsible for shooting the fish that will be eaten, are tiny hydras hanging on tentacles, with poisoned barbs at the ready.

As we look up the food chain to larger animals, nowhere do we find such an elaborate means of projecting venom. There is a species of cobra in Africa that bares its fangs and spits venom with great accuracy at the eyes of an attacker, often causing blindness, but this is a defensive action only. The poison, even when it enters the eye directly, will not kill, and the cobra does not follow up its spitting with a counterattack. It runs away.

Most snakes, of course, do not waste their venom spitting it away (nor does the spitting cobra, when it is on the offensive). Some venomous snakes deliver the poison by dribbling it down grooved back teeth, but most inject it through hollow fangs in the front of the mouth that work like a doctor's needle. The poison that fills them comes from glands inside the head. Usually the fangs fold back inside the upper jaw when the mouth is closed, but can be sprung into lethal position in an instant. These elongated teeth are fierce-looking, and no doubt their piercing stab is painful to an animal that is bitten, but of course they are not the snake's true weapon. The venom is.

Snake venom comes in two flavors. One attacks the blood, ruining its ability to clot; the other attacks the nerves, ruining the ability of muscles to obey commands from the brain and central nervous system. Each species of snake has one kind or the other, at a certain degree of potency. Sometimes, though, the chemical composition of one snake's venom in *one* habitat differs from that of the same kind of snake in *another* habitat; the snake seems to evolve the trick of producing a poison that will work most efficiently on the specific prey animals that surround it.

Originally, snake venom was merely a terrific digestive juice that allowed the small-mouthed, armless reptile to soften its meat before eating. Snakes really needed the predigestive step: they can't even *chew*. This is also how the venom of spiders and insects works—they inject their prey and essentially digest it outside their bodies, sucking it in only when the digestive acids have done their stuff. But somehow in the course of evolution, the snakes intensified the chemistry of their glands, and instead of enzymes and acids they sported venom created explicitly to kill. When a poisonous

*This is the last thing seen by many a desert rat
(a red diamondback rattler).*

snake bites an animal, it holds on long enough to squirt its whole supply of poison into the wound, and then it usually lets go. The terrified (and doomed) animal dashes away with hope of survival ("Hey, that wasn't so bad! I'm gonna make it!"). The snake, which knows better, follows at a leisurely pace. Before long it comes upon the prey, dead in its tracks, and swallows the animal whole.

Several venomous snakes are deadly to man, but most are not—the poison will cause pain, swelling, sickness, but not death. It is created to kill smaller, weaker animals; it is no more meant for us than a cow's milk is. Likewise, almost any snake will avoid biting humans, unless it is threatened or touched. Some of the snakes with the nastiest venom—kraits and sea snakes, for example—won't bite even when roughly handled. Like every other predator, the snake kills in order to eat. A king cobra that lives by eating other snakes knows very well it cannot swallow and digest a two-hundred-pound man, as does a rattlesnake that prefers gerbils and kangaroo rats.

Aside from snakes, only two other vertebrates orally inject venom: one lizard (the gila monster) and one mammal (a tiny but fierce shrew). Certainly we do not think of most birds, fish, and mammals working by such an indirect method as toxins that kill from within. We think of them biting, clawing, and bashing, more or less face-to-face, in simple combat. For many of the predators in the higher order, the actual kill *is* a mere matter of force, supplemented by something sharp (a beak, a tooth). This is, unfortunately, not the most biologically fascinating process; brute violence does not show nature's creatures at their clever (or even most athletic) best. A six-hundred-pound tiger jumps on a skinny impala's back and breaks its neck—big deal. An owl with a five-foot wingspan grabs a one-ounce mouse and crushes it with a squeeze of the talon (yawn). A twenty-foot shark bites a twenty-inch sunfish in half. So what?

The more interesting predatory features to reveal themselves in the actions that lead up to the kill or follow it. There is nothing especially wonderful, for example, in the triumph of the archerfish over a much tinier insect flopping on the surface of a stream. A quick bite, and that's that. But how the insect *got* to the water is amazing, and for this the fish deserves a prize. When the archerfish spots a bug on a leaf above its pond or stream, it cruises quietly beneath it, sticks a blowgun just above the surface of the water, and with a single dart shoots the bug down, PTING! (splash) MUNCH. The military precision of the attack is absolutely perfect. It's all

PTING! PTING! One of those drops shot by the archerfish will knock the bug off its perch and into the water, where it will flop helplessly until the fish gobbles it up.

the more amazing because the blowgun is not a bamboo pipe or steel straw—it's a tight little tube formed when the fish presses its thin tongue against a long groove in the roof of its mouth; and the ammo is not a poisoned dart or thorn—it's a droplet of water, released under great pressure.

Several other water creatures show physical ingenuity in the way they get prey close enough to eat. Anglerfish and alligator snapping turtles draw smaller critters near by jiggling a lure that attracts them, just as a human fisherman might. The difference is that the lures are part of these animals' bodies. In the case of the anglerfish, it's the dorsal fin's frontmost spine, which migrates toward the mouth as the fish matures, and develops a wormlike appendage that can be flicked at the end of a translucent "pole" as the fish lies still, camouflaged among the rocks on the sea bottom. The fish that drifts by to investigate will suddenly find itself sucked by an irresistible undertow, as the big predator creates a vacuum with its gills and pops its huge mouth open. This saves the anglerfish the trouble of lunging forward.

The snapping turtle's lure dangles from an even deadlier place: inside its mouth. The pink, wormish lure is nothing other than the tip of its tongue. The snapper's body, like the anglerfish's, looks a lot like a rock on the bottom. But to make its *open mouth* look like a small rocky nook requires a patience and immobility that would tax most creatures beyond endurance.

The alligator snapping turtle will hold its breath and act like a rock on the bottom of a pond for hours, moving only the pink nub on the tip of its tongue as a lure to curious or hungry fish.

Though we usually think of camouflage as a defense *against* predation, looking like a part of the scenery is the key to the kill for a lot of predators. Some look like parts of the plants favored by insects they prey upon: crab spiders that mimic the colors of flower petals and creep up on bees collecting pollen in the center of the bloom, walkingstick insects and praying mantises that appear to be twigs or leaves. Some look and behave like other creatures: the shrike that resembles the harmless mockingbird and shows its killer's colors only when unsuspecting birds and mammals have let it in among them. Some, such as the marvelous octopus, simply disappear.

The octopus pulls the greatest vanishing act in the world. A shellfish without a shell—free, therefore, from any kind of rigid structure—it can easily change its shape, to look like a rock, a piece of coral, a sea plant, or to fill in a crack and be nothing more than a smooth surface. More astonishing, though, is its ability to change color. It can mimic exactly the

Camouflage helps the ambusher. This crab spider blends in with the yellow petals until a bee loses itself among the pollen at the flower's center.

hue and pattern of a far broader range of backgrounds than other color-tricksters such as the chameleon; instantly the octopus can go from white to gray to brown to purple to aqua to red, in any order or combination, throwing in irregular patches of polka dots of contrasting colors for effect. The colors come from pigment sacs in the skin, controlled at will, guided by the very sensitive, analytical vision of the creature's single eye. Like the snapping turtle, the octopus puts its camouflage to work by waiting patiently for prey to come near. When a crab or fish is close enough, the octopus grabs it with a few of its eight sucker-covered legs, gives it a venomous bite with a birdlike beak situated at the hub of the legs, and eats it.

The suckers on the arms of octopuses and squids are fabulous predatory tools; they are little dishes of tough flesh that work exactly like the ends

Alert among the leaves it mimics, this Asian horned frog is ready to pounce on unsuspecting prey.

The octopus is the grand master of predatory camouflage, not limited to matching one background (like the horned frog), but able to change at will to whatever its single eye perceives.

Fish are hard to hold, so a squid's sucker-covered tentacles come in handy.

of suction darts, but with much greater strength. Sperm whales, which wage constant war against giant squids in the deeps of the oceans, are often found bearing great circular scars as much as three feet in diameter, from the pressure of a squid's sucker.

Many animals possess great, simple killing tools—powerful jaws (crocodiles, snakes, hyenas), sharp claws (hawks, crabs, cats), stingers (wasps, jellyfish, scorpions), and others. Sometimes the creature's entire body is its killing tool: the constrictor snakes coil around their victims and squeeze the air out of them. No tool is stranger, or more complex, however, than the one used by certain catfish and eels to kill their prey: electricity.

The electric catfish, found in the Nile and Congo river basins, and the electric eel, in South America, need only touch a fish or frog to overpower it. They sidle up close and turn on the juice—up to 500 volts—and ZAP! their meal is ready to eat. Sometimes the shock kills the prey; sometimes it only stuns it, but the result is the same. The electricity is produced by very specialized plates of muscle fiber that generate electric fields, store

The shrike's great weapon is its disguise. Except for that mean little hook to the upper beak, it looks like just another songbird, harmless as a mockingbird or towhee. Once the tweety birds let it join them, however, the shrike betrays their trust with a quick kill.

power like batteries, and project shocks like cattle prods. However, the catfish and eel have come to recognize that electricity has many uses: they also rely on it for a kind of radar, to navigate and find prey that show up as blips in their electric fields. One day, perhaps, they will develop interior microwaves.

When we look at the gifts predators have been given, coupled with advantages of size and strength, we may fail to understand why their struggle to feed themselves is so difficult. Why does the tiger or the anaconda have any trouble at all? How in the world do the prey animals continue to hold their own in the endless warfare, against such odds?

Well, no matter what tools a predator is given, from silk glands to poison to electricity, the animal still has to put them into use, often with skill, wits, and courage. Predators possess these necessary qualities; they get them from the primal urge to eat. But the urge to eat is matched in all of its force by an opposite determination, which can inspire just as much skill, intelligence, and courage: this is the urge not to be *eaten*.

The badger, looking rather easy to elude aboveground, hopes to drive his prey to digging—for he knows no creature can match his genius for speedy excavation.

ZOT! The chameleon's tongue is quicker—and can reach farther—than any claw, tooth, or tentacle.

Staying Alive

In the old days of baseball there was a green young pitcher with a great fastball. He blew it past hitter after hitter, but he kept getting into trouble because he thought his curve was equally fabulous. However, the hitters got a piece of the slower curve and put a lot of balls in play, some of which fell for hits and brought in a run here and there. The rookie's pitching coach tried to convince him to forget the curve for now and stick with the fastball. The kid resisted; the coach insisted; finally, the youngster demanded an explanation. "Why should I throw only my fastball?"

"It's simple," said the coach. "They cannot hit what they cannot see."

The parent of a young animal facing a lifetime of being chased as prey might adapt the coach's slogan: "They cannot eat what they cannot find." For when it comes to staying out of a predator's stomach, nothing is safer than the strategy of hiding. Even when a pursued animal's initial instinct is to fight back, it's probably better to flee: most prey are overmatched by the things that attack them.

It would seem that hiding would be easy. Predators are usually larger than their prey, and the world is full of cracks and crannies into which a smaller creature could scoot without fear of being followed. True, there are animals that pretend to mount a chase entirely so the prey *will* try to hide. The badger, for example, runs after a digging animal like a mole or a gopher, but it doesn't really want to do battle on its feet aboveground; it wants to frighten the critter into burrowing. The badger knows it digs better than anyone, so when the mole dives, the badger chuckles and burrows right after it, but faster. The escape hole becomes a grave. Some cats and bears will hope to drive prey up a tree for the same reason.

Most of the time, though, the prey animal's size gives it access to spots too tight for pursuit. An owl can't claw a field mouse out of a chink in a pile of rocks; a grouper can't suck a shrimp out of a groove in some brain coral. The trouble is, the prey animal also needs to eat, and to find food it must come out into the world and hunt or forage, just like the tough guys above it in the food chain.

Unless it is a termite. Termites seem to have the safest possible diet: they eat wood, and can spend their feeding time secretly beneath the bark of trees and fallen logs without ever showing their faces. But nature has

Almost any crack in the coral is big enough to hide a whole family of shrimp.

made these antlike insects the most popular prey on the land, giving a huge number of predators a taste for them, and providing quite a few of those predators with weapons designed exclusively to get at them. Wood-peckers, for example, have keen hearing to listen for the sound of termite jaws chewing beneath bark, beaks for breaking into the wood, and long sticky tongues for hauling the insects out by the dozen. Anteaters, aard-varks, pangolins, and other mammals have similar tongues, just so they, too, can eat termites.

If you have to be out and active, you want to be as subtle as you can about it. Camouflage helps. Animals have developed protective coloration with amazing sophistication. Sometimes it is as if the species that preys

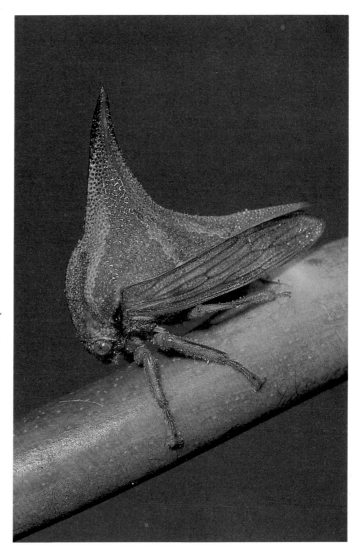

The thorn bug is doubly safe: it not only looks like part of the plant—it looks like a part of the plant that is both painful and inedible.

upon them had been consulted. ("Now, Mr. Towhee, we know you like spiders; but what do you find the most repellent thing, something you would never consider eating? Your own excrement? Very well." *Voilà:* there is a kind of spider that looks just like bird droppings.) In other cases, the camouflage works specifically against the optic capabilities of the particular predator—as with insects whose colors combine into a pattern matching their background so birds with a certain color blindness can't see them, or young shore birds whose mottled patterns disappear against the sand and rocks only because the gulls that would eat them have poor depth perception.

Defensive mimicry is much like camouflage, but instead of just blending in with the background, the animal imitates another (less edible) form of life. Many crustaceans in the sea resemble plants and flowers; so does a Brazilian spider, complete with pink-and-white flanges that look like petals. Mimicry of another *animal* usually requires not just coloration, but behavior as well. There is a species of grasshopper in Argentina that manages to copy the colors of a nasty wasp. But looking like a wasp won't fool anyone unless you also *act* like a wasp. So the grasshopper, if approached too closely, hunches up and curls its abdomen exactly as if it were going to deliver a sting.

Camouflage is not the only kind of protective coloration. Some animals depend on creating an optical illusion. A zebra, for example, is certainly not patterned to blend in with the unbroken plains of brown and green where it lives. In fact, it may seem at first that nature wanted to set the zebra up for the kill, to make it especially visible to the lions that hunt it. But sighting an animal is not the same as killing it, and the zebra's flashy stripes come to its aid closer to the moment of truth, when a lion closing in for the kill misjudges the distance between itself and the jumble of stripes. Thinking it is closer than it is, the lion bolts from cover too early, allowing the zebras an extra couple of seconds to scatter. Zebras will stand sideways to a lion spotted in the distance, angling their rumps (which have wider stripes than the rest of their bodies) toward or away from it, so that the lion is confused in its effort to interpret the zany planes of stripes.

When cottontail rabbits and white-tailed deer turn and flee from predators, they offer a clear marking to chase. "Follow me!" the white tails seem to say as they bound through the grass or forest. But the predator who focuses on the tail loses his visual grasp on the rest of the animal, so that when the white flash disappears—as the rabbit stops suddenly and

Zebras strike a pose that destroys their predator's depth perception. A lion could get a headache just figuring out how many there are and which one is closest.

sits, or the deer halts in a thicket and turns its rump away—he finds himself bedazzled and lost.

Some animals have the power or speed to get away without visual trickery. Most of the hooved mammals preyed upon by cats in Africa can outrun, say, the lion after a very few seconds; the cats need to make their kill in the first burst of speed and surprise, because they cannot sustain a long chase. Even the fastest land animal, the cheetah, runs out of zip after a hundred yards or so. It is true that no bird can outfly the diving peregrine falcon, no matter what; but this is probably the only predator that cannot be outraced in one way or another.

If an animal is not especially speedy, it can escape by being more of a master of its surroundings than its pursuer. The hooved mammals that live high in the mountains—certain antelopes, sheep, and goats—can flee predators simply by climbing, patiently and deliberately, to places impos-

sible for others to reach. These animals have split hooves, the halves of
which can be squeezed together over even tiny protrusions from the sur-
faces of rocks; it's like walking on four sets of powerful tongs. Unfortunately, these marvelous climbers are not chased only by terrestrial animals
more clumsy-footed than they are; danger can also come from above even
the highest climber. At the very moment a chamois looks smugly down
on a lynx from a safe peak, an eagle may swoop by with a buffeting rush
of wings and wind, knocking the chamois off balance into a thousand-
foot fall. The eagle will then beat the lynx to the body and register his
superiority in all matters of height by eating the presumptuous hooved
climber.

Well, if predators can dash from one environment into another to snatch
some prey, then a hunted animal can, too. If you can't get away from a
land-based killer on land, or a flying killer in the air, or a sea killer in the
water, the next thing to try is changing the scene suddenly, the way birds

*This wily crab defends itself by waving a stinging sea anemone in each claw when-
ever a predator comes too close.*

and insects pop into the air when threatened on the ground. But everybody expects a bird or a beetle to fly. Some animals add the element of surprise to their escape. The basilisk lizard, when pursued on the ground by a predator, rises onto its rear legs and dashes away with a two-legged technique worthy of the Olympics, which is unusual enough. What *really* shocks the snake or monkey chasing the basilisk is its disdain for the boundary between land and water: the lizard doesn't bother to slow down when it crosses a riverbank or the edge of a pond, but just keeps on running right onto the water. Its leg speed and technique allow it to sprint right on the liquid surface, and if it can scoot across without being spotted by a bass or catfish, it will surely have left its pursuer on the other side.

Amphibians, of course, can leave the land for the water, or vice versa, without discomfort; so can some of the reptiles, especially turtles and snakes. Unfortunately, there are predators in these groups, too, and they can follow their prey right through such ploys. At such times—when you have changed color, run as fast as you can, left the land for the water or the water for the air, and still the big-toothed meanie is hot on your heels— the only trick left may be to turn and fight. This is gutsy, honorable, and proud. In most cases, it is also death. A predator is *delighted* to get a piece of prey into some one-on-one.

At least, the predator is *usually* delighted. Sometimes it is in for a nasty shock or two. Quite a few prey animals are equipped with weapons they whip out only in the last resort ("All right, big guy—you asked for it!"). In the case of a wasp or venomous snake, the defensive weapon is the same thing normally used offensively, to kill prey, but in other creatures the hardware is activated only for self-protection. Several fish sport venomous spines that can deliver a fatal sting, but only when the fish is touched. A lionfish or stingray would prefer that everyone simply left it alone. The lionfish's appearance makes this clear: its flashy swordlike fins look every bit as dangerous as they are, and when threatened, the fish tries its best to warn the attacker off, rippling the blades with proud menace as if to proclaim DON'T MESS WITH ME. The same can be said of the tropical frogs that secrete fierce poisons through glands in the skin of their backs and advertise their toxicity by being brilliantly colored. An association of bright color with edible danger has evolved as a kind of signal code between predators and prey: almost all insects with defensive stings are sharply colored, as are these frogs and some of the venomous fish. Evolution made a bold decision somewhere along the line to rob these creatures of any camouflage, and thus to sacrifice a few generations to the birds and snakes and sharks that spot them so easily. But after a few

Sometimes the best defense is a good offense. This gutsy water shrew has grabbed the leg of a frog five times its weight; it will suck the bewildered amphibian out and kill it savagely. Animals, such as frogs, that regularly eat mice will avoid these fierce midgets. Once a biologist tried to feed a shrew to his pet rattlesnake in its cage; the next morning the snake was dead and the shrew was ready for more.

generations of predators learned that pain or death followed a colorful frog lunch or bee snack, the highly visible distinction became an excellent defense—good for the species, in the long run, if not for the individuals who gave their lives to prove their descendants would be dangerous to eat.

Venom is not the only chemical weapon used for defense. Quite a few creatures would rather stink than sting. Skunks, aardwolves, civets, and other mammals, as well as some insects, emit foul odors produced by glandular secretions. In most cases, the odor is something between a warning and an appetite deterrent, and it is emitted from the animal only as a scent. But in the case of the skunk, the foul perfume is sprayed forcefully into the face of the attacker, and from that point on, the odor is by no means limited to the skunk itself. The chemical is harsh in its makeup—

Like many poisonous animals, the "poison arrow frog" (from which rain-forest natives collect venom for their weapons) advertises its toxicity instead of hiding: Take a good look, *these colors say,* you know I'm bad.

it burns the skin and may cause temporary blindness, though the smell will last quite a bit longer. A predator that depends on its ability to sneak up on prey will find its hunting a bit hampered during the days it bears the skunk's stink, and it will probably never strike again at anything black with white stripes. In many places, the skunk is the boldest land creature in an environment where other small mammals skulk around in constant fear; one naturalist in Argentina has seen jaguars flee from it at the slightest warning twitch of the tail.

Squids and octopuses squirt stuff at their attackers, too, but it is neither toxic nor especially malodorous—merely dark, opaque, thick, and profuse. This ink spreads like a growing purplish cloud in the water around the predator, obscuring both the vision and sense of smell, while the octopus or squid darts away. A species of trunkfish uses the nature of water to spread a fluid in the same way, but with deadlier effect: it can secrete a toxin that kills predacious fish (or, for that matter, perfectly harmless ones) in the water around it.

The most horrific and primitive casting of fluid must be that of the horned toad. This thorny-looking but docile little lizard, which feeds on ants, has the alarming ability to shoot drops of its own blood from its eyes. And I mean *shoot*: just to make it clear that the gore is not the result of some injury—which would invite a predator closer—the horned toad sprays its hemoglobin with gusto, some six to eight feet outward.

The bombardier beetle is a chemist at the other extreme; where the horned toad squirts a natural bodily fluid for a simple effect, the bombardier concocts a series of chemical compounds to produce a sort of fireworks display. The chemicals are mixed separately and shot out all at once. When they reach the air, they explode with a loud, hot pop, just far enough from the beetle that it is not injured. The bombardier can mix its dynamite and eject it in an instant; often it will do so just as the tongue of a frog is arching toward it. The startled frog will remember, and think twice before trying to get another such bug into its mouth.

From stripes and spots to chemistry, defending oneself against predators can get pretty complicated. Probably the best defense has nothing to do with coloration, or speed, or fighting skill, or toxicity. The best way to stay uneaten may be simpy to taste bad, like the monarch butterfly and koala, which boastfully eat plants other creatures know to be toxic. Or perhaps the best way is to be covered with inedible spikes and plates, like the porcupine or pangolin. But even these defenses are not foolproof. Nature has a crafty way of placing in the world some creature somewhere that

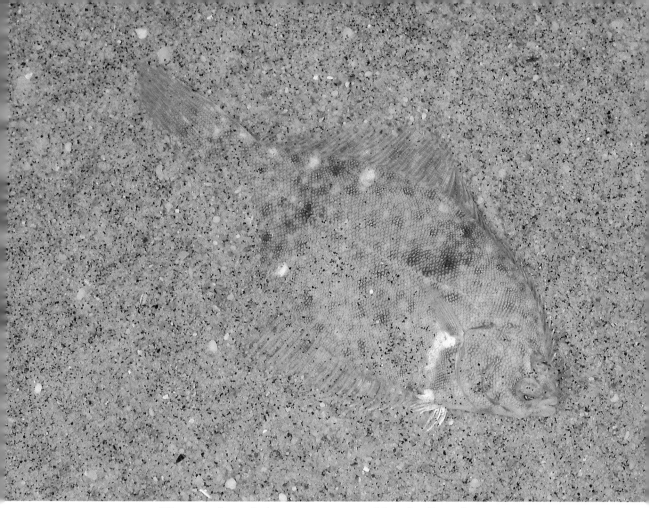

The art of not being seen, mastered by the flounder.

finds even the most repellent food yummy, or a creative hunter like the weaselly fisher, which takes weird pleasure in its difficult specialty, hunting porcupines. The fact is, almost everybody has a predator. It seems to make the whole system work, one animal feeding on another, then being eaten itself. The threat of death inspires strategies of behavior and reproduction that ultimately make the species stronger, by allowing the weaker, less crafty individuals to be eaten. A wolf may be a threat to a moose, but it is by no means a threat to the moose species, in the balance of nature.

To wipe out a species of animal today takes a very special kind of killer, operating outside the natural boundaries of an environment, skewing the laws that keep nature in balance. There is only one such killer, but he is clever, and, more and more, he is everywhere.

The Killer Outside

In truth, human beings are pretty wimpy predators. Very few hunt all the meat they eat, unless you call looking at newspaper ads to find which store has the cheapest chuck steak a *hunt*. Animals are killed, and their meat is eaten—but the distance between these two acts is too great to allow us the title of predator.

When we do hunt, we do it with a daunting technology that preempts any kind of defensive strategy. What duck or deer can fight back against a shotgun or rifle? But this isn't really important—humans are not cheating just because we've invented techno-weapons. If the wolverine (supposedly already the fiercest fighter in the world) could suddenly get some poison glands and needle fangs (or an Uzi, for that matter), it would gladly use them. When one hunter is going after one animal, there's no such thing as overkill—there's just a kill, and whether it takes place through twelve bullets shot in 1.6 seconds by a semiautomatic rifle or through a single arrow shot from a wooden bow, the difference matters only to the hunter.

Nevertheless, humans represent the wildest danger ever encountered by animals, as absolutely crazy killers whose feats of slaughter would overmatch all the Siberian tigers and rattlesnakes that ever existed. The reason is this: humans do not always kill within nature's balance of power. We tend to stand *outside* the system, shooting in.

The dangers we pose have nothing to do with hunting for food. On the contrary, human populations that depend on hunting have usually been "cleansing" predators, like wolves or pumas, taking the easier kills: the slower, older, weaker animals in the herd of buffalo, caribou, antelope, or deer. This keeps the breeding population strong, as the healthier animals stay alive to produce healthy offspring. The general health of the herd is something the dependent human hunters cannot help but notice. Many Native American peoples in North and South America followed codes of hunting behavior that served a scientific purpose as much as a moral one: there were efficiency and conservation in the way they selected, killed, and ate their prey. In Africa today there are several tribes of hunters who have integrated themselves into the food chain, coexisting with lions at the top in an unusually close competition. These hunters seek the same game the lions do, and although each group of predators sticks to a rigorous schedule (men have the day, lions the night), sometimes they overlap and must argue over a kill. Sometimes, too, one group having hard times must poach an animal just killed by the other. The lions and men work these things out by voice and gesture and intuition, with respect and fearlessness. They simply never attack each other.

When we hear of humans and lions sharing a hunting ground and stealing food from each other face-to-face, we marvel that the lions do not slaughter the men. After all, lions are so vicious, instinctive, and crude—how could they restrain themselves? But really it would be more realistic to marvel at the fact that the *humans* do not wipe out the *lions*. That is how such competitions usually turn out, especially now that we have such a hot arsenal.

When a group of sheep ranchers in the southwestern United States declares war on coyotes and pays bounty hunters to round them up and shoot them by the thousands, or midwestern farmers decide to obliterate a population of beetles, it's as if a spectator at a football game suddenly started shooting one player, then another, then another. Pretty soon the intricate game breaks down. When the center is gone, no one can hike the ball accurately to the punter. Without a tight end, a team's running back is suddenly open to a big rush from the outside linebackers. As the wide receivers disappear, passes sail downfield and bounce, untouched, on empty ground. One team suffers greater losses, and the other team roars over it in a stomp.

Possibly the game could go on if substitutes could take the place of fallen players. But in the contact sport of the food chain, there are no substitutions. Eliminate or weaken a species that feeds others—and feeds *upon* others—and you change everything forever.

Humans tend to look at creatures individually, out of context, and pass judgment quickly on an animal's fate. Chinese farmers see sparrows eating their crop seed, so the government launches a campaign to wipe out sparrows. Shortly thereafter, the crops are munched to the ground by an unprecedented plague of caterpillars, year after year. Guess what the sparrows used to eat, far more than they ate grain? That's right—caterpillars, up to a hundred a day per bird. So now the government has to buy expensive chemical pesticides to replace the natural ones they eliminated. The caterpillars grow resistant to the poisons, but other creatures die from them. Crustaceans are poisoned by the toxins that leach from cropland into the rivers, and the fish that feed on them decline in numbers, drastically reducing the fishing economy and food supply of many villages. The shells of eggs laid by hawks that eat polluted fish are fatally softened by chemical residues, and the hawks begin to die out. Suddenly the vermin they used to prey upon—rats, especially—start increasing, unchecked. The rats eat into the granaries and gobble up ten thousand times the grain that was originally begrudged the sparrows. And on and on and on it goes, more and more complicated and expensive and dangerous. All because we decided to interfere.

There are ways of using our ingenuity and technology to control pests without going too far outside the laws of nature. For example, scientists can breed pest insects in laboratories, but alter their genes so that they are unable to reproduce. When these insects are released into a surging population of their own kind, they negate the fertility of the "natural" population, and the birth rate plummets. On a smaller scale, a gardener whose plants are infested with caterpillars and beetles may hunt in wild fields for egg cases of the praying mantis and "plant" them around the garden. When the mantises hatch, they leap upon the bloated hordes of insects and restore a kind of balance.

Sometimes, though, the introduction of a predator backfires as terribly as the subtraction of one. On some tropical islands, biologists who wanted to control the population of snakes imported breeding pairs of the mon-

Caterpillars can wreck a garden or crop, but only until the local songbirds find out about them.

Are rats stealing your grain? A few corn snakes will take care of the problem, no mess, no venom, no fee.

goose, a small deft mammal that is a great snake-killer. Soon the islands were *completely* snake-free. But now they were pretty crowded with mongooses, which had bred and thrived while the snake feast was at its peak. The mongooses discovered that they liked bird eggs and chicks, so the islands started losing birds (and gaining insects, which the birds used to eat). It would have been better if the mongooses had transferred their attention from the deleted snakes to the rodents the snakes had preyed upon—but they opted for the bird eggs, and the rodents, liberated from their ancient reptile enemies, proliferated madly.

Introducing a predator into an environment is not always intentional. In the sea trade, a ship stopping at a port carries all sorts of unknown stowaways that may be left behind onshore or in the waters to establish populations that are not native to a particular spot. This is especially dangerous on islands, where native animals have had the security of isolation for centuries, in which they learned exactly which animals they could eat and which would eat them. A new predator in such a place finds itself in paradise, for the native animals tend not to fear it, so established are their habits of security.

On the island of Guam there are no native snakes, except for one wormish blind one that eats only soil insects. Because of this, the birds of the island did not evolve any of the usual defenses adopted by birds sharing a jungle with the egg-eating reptiles: they didn't hide their nests in cavities, or hang them from the ends of tree limbs as the weavers do, or use a communal sentry system to keep a sharp lookout. They built them out in the open, because they had nothing to fear. But one day a ship docked in port, and—probably by accident, but perhaps because someone thought it might be fun or useful to see what would happen—a couple of snakes from the Philippines crept off the ship and into the jungle. There they found all the eggs they could wish for, displayed openly for the taking. They ate, and bred, and ate, and bred . . . and still the birds had no instinct about what to do about these odd new creatures that kept sucking down their offspring. The result: Guam used to be an island of birds. Now it is an island of snakes. There are only a few hundred birds, of any kind, left on the place, but there are perhaps ten thousand of these Philippine snakes *per square mile.* (Of course, the animals that preyed upon the snakes in their native environment are absent from Guam, too.) Six bird species that existed on Guam and nowhere else in the world have pretty much vanished forever. This has all happened in forty years or so—far too short a time for evolution to catch up and instill some defensive tactics in the birds.

Not all applications of science to animals are destructive, of course. The technologies of death have inspired technologies of conservation, dedicated to preserving threatened animals and restoring them to their place in protected natural environments. Sometimes the people who helped start the problem recognize their errors and try to help reverse it (duck hunters establishing conservative limits on certain species, farmers converting their pesticide-fried fields to organic production). Sometimes a very limited success is achieved (the comeback of the alligator in Florida). But reinserting a fragile population of animals into a guarded territory is at best a symbolic victory, little more than making a glorified sort of zoo. When a predator had been yanked out of a slot in its environment, the slot closes, and animals reintroduced to the wild will usually find that they cannot force their way back in.

Such are the risks when humans somehow mess with the balance of predator and prey. Yet here we are, atop the world's hierachy of manip-

Gardeners search for praying mantis egg cases to place among their plants, so the hatchlings will establish a reign of terror for pests.

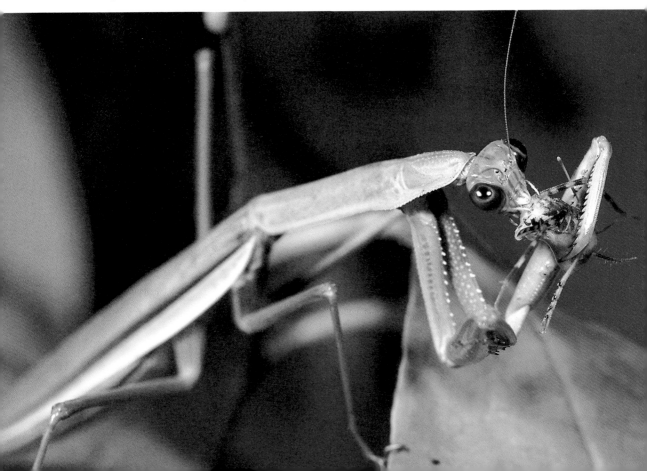

Is the polar bear—roaming a barren land of ice and scrub—a sign of what predators in richer biospheres are facing in the future?

ulative intelligence, blessed with a vision for technologies that we've grown to depend on, with needs of our own. Certainly it's not really *our* fault that we stand outside the food chain. What are we supposed to do, go back to the days of hunting woolly mammoths for food? With wooden clubs? From the earliest eras, humans *had* to develop tools and the wits to use them, in order to eat; we did not have the raw power and fierce physical tools of the animals with which we often competed for prey. It could be argued that the ability to throw some dynamite in a pond and collect two thousand dead bass from the surface five seconds later is exactly what evolution has prepared us for—that such killing is our right and our destiny. Why should we have to restrain ourselves? After all, we, too, get hungry.

But being so smart gives us the ability to think beyond the immediate satisfaction of hunger. It is a rare privilege to be free from the need to hunt all day long; perhaps this privilege gives us the responsibility to take more time before we act, to consider nature's design and to understand it. Not just to *alter* it—but to understand it.

Glossary

Adaptive—An animal adjusting to conditions that have changed is adaptive; it meets unexpected demands by using its instincts, intelligence, and physical abilities to find or create what it needs to survive.

Amphibian—This is a class of vertebrate animals that can live in water and on land: frogs, toads, salamanders, newts, and wormlike caecilians. Most amphibians have very simple lungs, as adults, and breathe largely through their moist skin. They are cold-blooded—their body temperature depends on the temperature outside, instead of being regulated from within, as it is in warm-blooded animals. Scientists believe amphibian species used to be fish; millions of years ago their ancestors left the water and, over the course of generations, developed lungs and legs.

Bird—If an animal has feathers, it belongs to this class of vertebrates. Most birds fly and use this gift in all parts of their lives, but some do not; one (the kiwi) does not really even have wings. Birds live very intensely—they are amazingly quick and sensitive, and their hearts beat up to ten times faster than ours do (they are warm-blooded, with a body temperature as much as 12 degrees higher than humans). In North America, we see more of them in the spring and summer because many species fly south to spend the winter in warmer parts of the world (this is called migration).

Crustaceans—These are sometimes called "insects of the sea"; they are in fact in the same large grouping (the arthropods) as insects and spiders. Crustaceans are a primary food for many marine animals and seem to be everywhere in both salt and fresh water. The 26,000 species include many tiny and simple creatures, such as brine shrimp, and other, more complex animals such as crabs and lobsters. Some of the crabs live on land as well as in water; the wood louse is a crustacean that is entirely terrestrial.

Environment—An environment is the physical surroundings—everything from rocks and air to plants and animals—among which an animal lives. The term is used in two ways. First, scientists define several specific types of environment according to clearly distinct features (for example, salt marsh, old-growth forest, tundra); they try to figure out the very complex relationships between all of the living and non-living things found in that particular kind of place. (In general, they have learned that if one aspect of an environment is changed, even a little, then everything in it is affected. Nothing is unimportant.) Second, all of us—not just scientists—have begun to speak of the environment as the complicated physical wholeness of the earth: the air, water, minerals, and animals among which we must fit.

Evolution—This is a slow, orderly, progressive process through which animals change their physical makeup or behavior, as a means of thriving in a changing world. An evolutionary change takes a long time over many generations of a species (as contrasted with an adaptive change, which can take ten minutes in one individual). In general, animals are always competing—for space, food, and mates with whom they can reproduce their kind. When a certain feature leads one individual to succeed over others (a slightly longer neck that lets it reach higher leaves richer in nutrients; brighter feathers that attract stronger mates; a deeper, more aggressive roar that frightens trespassers; and so on), the animal eats and breeds more vigorously, passing that feature on to its offspring, which in turn succeed, in greater numbers, and pass it on—until the new feature becomes a universal fact of life for that kind of animal.

Host—A living thing upon (or inside) which another living thing lives is a host. Sometimes the intruder more or less borrows space, and doesn't especially bother the host; sometimes the intruder steals from the host (or even devours it from the inside) and weakens or kills it.

Innate—A capability or knowledge that an animal is born with is said to be innate. Such abilities can be physical or behavorial; usually they are a combination of both. A young gull, for example, is born with wings upon which feathers grow, and it will naturally put them to use in the maneuvers of flying.

Insect—This is the largest class of animals in the world: there are about 800,000 known species, and more are being discovered all the time. Insects are invertebrates—they have no skeletons inside. Their adult bodies are divided into three segments (head, thorax, abdomen); they all have six legs, two compound eyes, and a pair of antennae; most have one or two pairs of wings. They live in more places in the world than any other kind of creature, in larger populations. One well-known scientist said to another, "What would life be like if the insects took over the world?" The second scientist laughed and said, "My dear sir, haven't you noticed that they already have?"

Instincts—Animals are born with knowledge about how to live their kind of life. The interior mechanism that supplies this knowledge and puts it into involuntary use is instinct. Instinct gives spiders a mental blueprint for their webs; it makes newly hatched chickens hunker down and peep in terror when a V-shaped shadow passes by them, though they have never seen such a shadow cast by a real hawk (for that matter, neither have they seen what a hawk likes to do to a chicken). When we watch animals, we are amazed at how self-sufficient they are. Instincts are the reason they are able to take such detailed care of themselves, with so little study or instruction. But however exact instincts may be, they do not equip animals to solve unexpected problems with analytical intelligence; instincts are rigid instructions that can only be rigidly applied.

Invertebrates and vertebrates—Vertebrates are animals with backbones; invertebrates are animals without. More than 97 percent of the world's animals are invertebrates. Some of them are basically blobs (amoebas, jellyfish); some are soft but well formed (octopuses, earthworms), some have armored themselves with hard shells that give their bodies a tough definition (coral, lobsters, beetles). Perhaps because humans have backbones, we regard vertebrates as the higher life forms, despite their meager numbers. Mammals, birds, reptiles, amphibians, and fish all have spines; most have internal bony skeletons as well, though some have cartilage instead (sharks), and others have bony plates under the skin (trunkfish) or outside it (turtles).

Larva—Many animals are born with a body quite different from the one they will inhabit when they grow up. They go through one or more changes of shape and eventually attain their maturity. In the early stages the animals are called larvae. Often a larva bears no resemblance to its future form (caterpillar to butterfly, tadpole to frog). Larvae may also live in a different type of environment, follow a different diet, and thrive on a different life-style than the ones they will later settle into.

Mammal—We are pretty familiar with the vertebrate class called mammals; humans belong here, alongside dolphins, dogs, elephants, rats, whales, and lots of other warm-blooded creatures that give birth to living young (except for the oddball platypus and echidna, which lay eggs) and feed them milk. Mammals have one set of replacement teeth, fingernails or claws or hoofs, large brains, and hair. Most mammals walk the land on four legs, though quite a few have adapted to life in the water, some merely showing vestiges of their quadrupedal structure; only one group, the bats, can truly fly, though several other species spread flaps of skin and glide.

Organism—Something that lives.

Photosynthesis—All plants absorb light from the sun and convert it to chemical energy, which they use to make sugar from water and carbon dioxide from the atmosphere. This sugar can be stored and used for food. The process is called photosynthesis. This self-sufficient ability to manufacture food from scratch makes plant forms the absolute base of the food chain: they do not have to "catch" anything to eat—they need only sunshine and moisture.

Reptile—Snakes, lizards, crocodiles, turtles, and some wormlike things called amphisbaenians make up the class of vertebrates called reptiles. Almost all lay eggs that hatch into babies that look just like the adults (no larval stage). Reptiles are cold-blooded, and most have a scaly, dry skin.

Species—A type of animal that is different from others by reason of physical or geographical distinctions is said to be a species. It's simply a word for the units of individuality in the natural world. Sometimes the distinctions between one species and another are very slight, but generally they follow the animals' own rules of division, mainly expressed in breeding habits: creatures tend to mate within their own species.

Vertebrates—See "Invertebrates."

Acknowledgments
and Photo Credits

The Knowing Nature Books are inspired by the broad spirit of inquiry and richness of detail in the *Nature* television series. The books are original works, however, and their material is not derived from the *Nature* programs. Thanks to those who make the *Nature* series possible: George Page and David Heeley at Thirteen/WNET, New York, with the generous support of the American Gas Association, Siemens, and Canon.

Personal thanks to the people who have helped make the book: David Wolff, Lee Anne Martin, Margaret Ferguson, Elaine Chubb, David Reisman, and Licia Hurst.

Cover photo is of a Siberian tiger; photo on title page, an emerald tree boa with an opossum; photo on page 2, a red fox; photo on page 12, a polar bear; photo on page 28, a red-tailed hawk with diamondback rattler; photo on page 44, a snowshoe hare; and photo on page 56, an airplane spraying pesticides on a field of winter wheat.

Cover © John Chellman/Animals Animals
Title   © Jany Sauvanet, National Audubon Society Collection/Photo Researchers
PAGE
2   © Ken Cole/Animals Animals
4   © Sierra Leone, National Audubon Society Collection/Photo Researchers
5   © Michael Fogden/Animals Animals
6   © Preston Mafham/Animals Animals
7 (top)   © John S. Dunning, National Audubon Society Collection/Photo Researchers
7 (bottom)   © Michael Dick/Animals Animals
8   © Oxford Scientific Films/Animals Animals
9   © M. Reardon, National Audubon Society Collection/Photo Researchers

10   © Leonard Lee Rue III/Animals Animals

11   © Zig Leszczynski/Animals Animals

12   © Wayne Lankinen/Bruce Coleman Inc.

15 (top)   © Sean Morris, Oxford Scientific Films/Animals Animals

15 (bottom)   © Sean Morris, Oxford Scientific Films/Animals Animals

16   © J. H. Robinson, National Audubon Society Collection/Photo Researchers

17   © Kevin Jackson/Animals Animals

18   © Zig Leszczynski/Animals Animals

20   © J. Gregory Brown/Animals Animals

21   © Carl Roessler/Animals Animals

23   © Anup & Manoj Shah/Animals Animals

24   © Edgar T. Jones/Bruce Coleman Inc.

25   © Robert W. Hernandez, National Audubon Society Collection/Photo Researchers

26   © Joe & Carol McDonald/Animals Animals

28   © Len Rue, Jr., National Audubon Society Collection/Photo Researchers

30   © Stephen J. Krasemann, National Audubon Society Collection/Photo Researchers

31   © Tom Branch, National Audubon Society Collection/Photo Researchers

33   © Tom McHugh, National Audubon Society Collection/Photo Researchers

35   © G. I. Bernard/Animals Animals

36   © Cosmos Blank, National Audubon Society Collection/Photo Researchers

37   © Ken Brate, National Audubon Society Collection/Photo Researchers

38   © Zig Leszczynski/Animals Animals

39   © Zig Leszczynski/Animals Animals

40   © Bill Wood/Bruce Coleman Inc.

41   © Charles Palek/Animals Animals

42   © Stephen Dalton/Animals Animals

43   © E. R. Degginger/Animals Animals

44   © Leonard Lee Rue III/Animals Animals

46   © David Hall, National Audubon Society Collection/Photo Researchers

47   © James H. Carmichael, Jr., National Audubon Society Collection/Photo Researchers

49   © Terry G. Murphy/Animals Animals

50   © Kjell B. Sandved/Bruce Coleman Inc.

52   © Dwight R. Kuhn/Bruce Coleman Inc.

53   © Zig Leszczynski/Animals Animals

55   © Breck P. Kent/Animals Animals

56   © Junebug Clark/Photo Researchers

60 (top)   © Alastair Shay, Oxford Scientific Films/Animals Animals

60 (bottom)   © Richard R. Hansen, National Audubon Society Collection/Photo Researchers

61 (top)   © Stephen Dalton, National Audubon Society Collection/Photo Researchers

61 (bottom)   © Zig Leszczynski/Animals Animals

63   © Breck P. Kent/Animals Animals

64   © Johnny Johnson/Animals Animals

Index